BUILDING BLOCKS OF LIFE

OF

LIFE

HOW ESSENTIAL AMINO ACIDS
KEEP YOU YOUNG, VIBRANT, & HEALTHY

ROBERT R. WOLFE, PHD

The Building Blocks of Life
How Essential Amino Acids Keep You Young, Vibrant, & Healthy

Published by TriVita Press

Live Stronger at Any Age, Vol. 1

ISBN: 978-1-943157-46-4

Printed in the United States of America

To order go to:
www.trivita.com

LIVE STRONGER AT ANY AGE!

Preface

Essential amino acids are the fundamental building blocks of life as we know it, but there are specific instructions involved.

After decades of research and study, Robert R. Wolfe, PhD, found two factors (a unique ratio and balance of quantity) of essential amino acids that when consumed in supplement form, based on this specific formula, brought about great results, such as improved health, rapid healing, greater strength, and sharper mental capacity.

The life-changing benefits of this unique formula are what you are able to enjoy as a result of his efforts. The essential amino acid solution (EAAS) formula is what makes the MyoHealth products (in capsule, powder, or protein shake form) so incredibly effective.

This book shows the what, why, and how behind the EAAS formula.

Table of Contents

Foreword

I am not claiming to have discovered the fountain of youth, as anyone who makes that claim should be regarded with a good deal of skepticism.

We cannot stop the process of aging entirely … but there is a lot we can do to help our bodies perform at higher levels and be stronger and more mobile, regardless of our age.

The biggest boost to improving quality of life, I have come to believe, comes from essential amino acids. In fact, in the years to come, I expect to see essential amino acids become the most important nutritional supplements to impact human health and disease.

My interest in the nutritional role of these amino acids began when I was a 13-year-old freshman in a California high school with big hopes of making the varsity basketball team. I had the height (6'5") and the inspiration, but I had one major problem—I was a rail-thin 165 lbs.

When practice started, it became evident quite quickly that the older, stronger players could easily push me around the court. Disappointingly, I was relegated to the junior varsity team. Right then and there I vowed to build up my muscles and be much stronger over the next year so I would not suffer the same fate again my sophomore year.

Now, I intuitively understood the role of nutrition and had the concept that optimal nutrition would be crucial in my endeavor to gain muscle mass and strength, but I had no idea how to actually do it. Little did I know that essential amino acids held the key.

I was able to get strong enough to make the varsity team as a sophomore and eventually to have a successful career at college and, ultimately, to be drafted by the Warriors in the NBA, but I never weighed more than 190 lbs.

Hindsight is always 20/20. I know my basketball career would have benefited from my current understanding of muscle metabolism and the important role essential amino acids play in developing muscle mass and strength, but I had no idea back then.

While in college, I chose to make a career of exploring how to improve physical performance, overall health, and quality of life. I would do that through optimizing protein and amino acid nutrition.

Since then, I have spent almost five decades immersed in research on human physiology and metabolism. I have performed studies on Olympic swimmers and cyclists, and have worked to help people recover from major diseases and injury. As the Chief of Metabolic Research at the Shriners Hospital for Children at Harvard Medical School, I performed research on nutritional and metabolic support following severe injury that translated to guidelines for patient care which are now recommended in both the United States and Europe.

After 10 years at Harvard, I moved to a similar position at the University of Texas Medical Branch, in which I was able to expand my research to a variety of conditions, including sepsis, traumatic brain injury, cancer, and diabetes.

Over the next 25 years, I became increasingly aware of the importance of muscle in recovery from all of those conditions, and, in a broader context the importance of muscle in maintaining health and quality of life throughout the lifespan.

Consequently, I moved to the Donald W. Reynolds Institute on Aging at the University of Arkansas for Medical Sciences, in order to focus my work more directly on health issues related to aging and longevity. I became the Wormack Chair of Geriatrics and Director of the Center for Translational Research in Aging and Longevity. In this position, my main role was to perform research that will directly translate to the betterment of people's lives, where muscle health actually contributes to mobility and reduction of disease, adding up to a much-improved quality of life.

After I reached age 70, things I never thought of in my younger days were now of concern, such as my blood glucose and lipid levels, my diminishing strength and ability to do things that were normally easy to do, and recovery from surgeries (ankle, knee, and hip).

For me, the surgeries are a direct result of running thousands of miles. Being competitive and inquisitive, I continually pushed my own limits. I ran to and from work for many years, and trained incredibly hard for marathons. The extreme wear and tear on my body is the result, but I am thankful for the essential amino acid solution (EAAS) formula that has helped me recover from several surgeries and continue to live a mobile and active life.

The EAAS formula I created for healthy aging addresses the issues we all face ... and I have been using the formula myself. You could say I have been "practicing what I preach" for a few years and have found that gradually my clothes have started fitting better, I have more energy, and my blood chemistries have reached the levels of a young, healthy man. Following my hip replacement, my physical therapist repeatedly commented that he had never before had a client who recovered so quickly.

Though I have reaped incredible benefits from my EAAS formula, I recognize that my personal experiences are not a scientifically tested study ... but there are 24 human studies and over 20 million dollars of research behind this essential amino acid formula.

That means I am confident that you can expect to achieve similar results by making a long-term life commitment to the essential amino acid solution (EAAS) formula that will help you live stronger at any age.

Over the years, I have published almost 600 peer-reviewed research papers that have been cited by scientists in other articles more than 50,000 times. In the following pages of this book, as I discuss the what, why, and how of essential amino acids, I will show how the proper use of essential amino acids can improve your quality of life.

As a result, I fully expect to hear how your health, vitality, and quality of life are improved.

To your success and future!

—Robert R. Wolfe, PhD
Professor, Department of Geriatrics
Jane and Edward Warmack Chair in Nutritional Longevity
Director, Center for Translational Research in Aging and Longevity
Donald W. Reynolds Institute on Aging
The University of Arkansas for Medical Sciences

Introduction

The proper use of the essential amino acid solution (EAAS) formula, coupled with simple nutrition and exercise strategies, improves the health and functioning of everyone. That is a fact.

The intent of the EAAS formula is to augment your normal life pattern without requiring you to completely restructure the way you eat and live.

This means that you will not be overwhelmed with a mountain of foods or supplements to face each day. Similarly, recommendations for exercise will not require you to join a gym and devote hours every day to fitness and training.

Rather, the fundamental commitment asked of you is to incorporate a few changes into your existing dietary pattern and supplement usage. Doable? That's not too daunting nor does it need to be.

I do encourage you to engage in some level of physical activity. Find enjoyable activities that produce the maximum benefits for your muscles, and that will positively impact all areas of life.

While the nutritional tactics alone will provide benefits, the synergy between amino acids and physical exercise that you will learn in this book is so tremendous that even the most reluctant exerciser will be inspired to engage in some level of physical activity. You will be rewarded by the exercise, and, in combination with the nutritional program, the benefits will amplify. You will feel the positive effects of more vitality and muscle strength within a few short weeks!

Most importantly and fundamentally, the guidelines are designed to become a way of life, not just a crash diet or "detox" that you suffer through for two or three weeks. That is no fun, nor is it going to help anyone achieve lasting health.

I am confident you will start to feel the effects of the essential amino acid solution (EAAS) formula within a few weeks, but you will not feel the full impact of the program for two to three months. This is committing to give your body the balance of essential amino acids you need to build stronger muscles, and muscle mass to reduce the aging problem of muscle atrophy and the disease issues that go with it.

Regardless of your current physical and health status, I am confident that the scientifically validated essential amino acid solution (EAAS) formula will improve your physical and metabolic function, and that will lead to better physical health and quality of life.

CHAPTER ONE

What Are Amino Acids?

Amino acids in the body are the fundamental building blocks of life as we know it. These chemical compounds made of nitrogen, carbon, oxygen, hydrogen, and a handful of other elements join together to form protein, the material that forms the muscles, tendons, and organs in our bodies.

In addition, amino acids are the main ingredients of most of the biochemical components in our blood and cells that are necessary for life.

Simply put, everything is *built on* or *built out of* amino acids.

The function of amino acids includes the following tasks:

- building muscle and life-supporting tissues
- making the chemicals necessary for the brain and vital organs to function

- playing an important role in many metabolic pathways

Importantly, essential amino acids (EAAs) are the only macronutrients that are absolutely required in our diet. They not only keep us alive, they improve our metabolic function and physical capacity.

The Dietary Amino Acids

Technically, the term "amino acids" refers to the general chemical class and consists of more than 300 individual compounds. Only 20 of these amino acids occur in proteins in the body. These amino acids are "dietary amino acids" because they are also in the proteins that we eat.

> Amino acids affect our health and well-being in countless ways.

Of these 20 amino acids, 9 are considered essential and 11 are non-essential. It is vitally important to understand just how essential the 9 essential amino acids are! Without them, you will die, and the only way to get them is through the foods you eat. That certainly counts as essential!

The 9 essential amino acids (EAA) include the following:

> leucine, isoleucine, valine, methionine, phenylalanine, tryptophan, threonine, lysine, and histidine.

The 11 non-essential amino acids (NEAA) include the following:

> cysteine, glycine, glutamine, proline, arginine, tyrosine, alanine, aspartic acid, asparagine, glutamic acid, and serine.

It is important to know that regardless of whether an amino acid is essential or non-essential, each amino acid plays a part in protein synthesis.

Why We Need Dietary Amino Acids

The protein we eat is composed largely of amino acids. When we eat protein, it is digested in the stomach and intestine, and the amino acids are absorbed into the body.

Did you know that there are thousands of different proteins in your body, all with specific functions? Not counting water, those proteins compose about two-thirds of the mass of your body.

Regarding these proteins, it is vital to understand this important fact:

> Every protein in your body is in a constant state of breakdown and synthesis.

It is this balance between the rate of breakdown and the synthesis of a protein that determines whether you are gaining (anabolism) or losing (catabolism) protein. Most people are in a steady state, in which the overall synthesis of proteins balances out the breakdown of proteins.

> The fact that there are 9 different essential amino acids is why we need to eat a variety of protein food sources.

The breakdown of body proteins provides a steady supply of amino acids to produce new proteins. Even though there is a constant supply of amino acids coming from protein breakdown, we still need to get amino acids from the food we eat. This is because all of the amino acids that are released in the process of protein breakdown are not available for reincorporation into protein via protein synthesis.

While some amino acids (the non-essential amino acids) can be produced from other molecules in the body, other amino acids released as a consequence of protein breakdown are irreversibly degraded and excreted (as CO_2, urea, and ammonia).

Because we are losing protein and cannot make what we need (our body cannot produce the 9 essential amino acids), the only answer is to eat protein that contains the essential amino acids.

You drink water daily to meet your body's daily need for water. Similarly, your body needs amino acids on a daily basis to offset what your body uses, discards, or loses.

> Your body needs essential amino acids … on a daily basis.

By eating sufficient essential amino acids, which is best accomplished through a balanced approach (food plus the essential amino acid solution (EAAS) formula), the rate of protein synthesis can match or even exceed the rate of protein breakdown.

At that point, you have growth, strength, and restored vitality that leads to a greater quality of life as well as a reduction in risk of disease.

It is simply essential that you get your essential amino acids!

CHAPTER TWO

What Amino Acids Do for Your Body

As you know, we must obtain the necessary balance of essential amino acids from food sources or supplements since we do not have the ability to make them. Your body, on the other hand, can make or synthesize the non-essential amino acids.

When it comes to eating the non-essential amino acids, most people usually consume more than they need. There is no recommended dietary allowance (RDA) for the non-essential amino acids.

However, there is a very strong RDA for the essential amino acids. After all, it is the essential amino acids that control the rate of protein synthesis.

To make matters worse, people usually do NOT get sufficient amounts of essential amino acids, as it is very challenging to eat a diet in abundance and balance of the essential amino acids. The body can cope, or hobble along if you will, but it is not performing at its peak by any means.

> The requirements for dietary amino acids are even greater under stressful conditions.

Inevitably, the deficiency will eventually lead to accelerated muscle loss, which is very common with people 50 and above.

Each Amino Acid Plays a Part

Every amino acid is a structural component of protein. The principal role of dietary proteins is to provide the amino acids that serve as precursors for the production of new protein, to balance the amount that is lost daily through the process of breakdown.

Many amino acids play additional roles. For example, arginine plays a role in regulating blood flow and blood pressure as a precursor for the production of nitric oxide, which is the primary chemical responsible for dilating blood vessels, particularly in muscle. The essential amino acid leucine can activate the molecular pathways involved in the initiation of protein synthesis.

Every essential amino acid plays a unique and important role.

It is at this point that many people get in trouble and hurt their bodies.

They focus on a single task accomplished by a single essential amino acid and reason, "I want more of that, so I will take a supplement high in that essential amino acid."

What they do not know is that if they take a specific amino acid supplement, the adverse effects will likely outweigh the benefits sought.

Why?

A high-quality dietary protein contains a ratio of EAA to NEAA of about 55:45, which is similar to the makeup of human muscle.

Because the amino acids do not work independently of each other … they work together in unison.

What that means is this:

> Taking the essential amino acids in the correct proportions is just as important as taking the essential amino acids in the first place.

It is important to know that a supplement containing a high amount of one or several essential amino acids will not benefit the body. It cannot because the body dumps the excess.

Many of the non-essential amino acids are made in the liver.

Yes, avoiding a deficiency is important, but taking a dietary supplement that creates an excess of several amino acids causes greater harm. The intentional overdose of essential amino acids, because it puts the body out of balance, damages some of the other essential amino acids during the re-balancing efforts.

There is loss initially when all excess essential amino acids are dumped, and there is further loss in the effort it takes to bring the body back into balance.

The best way to approach optimal amino acid nutrition is with a balanced approach, which includes healthy foods plus the essential amino acid solution (EAAS) formula. Together, this brings strength, mobility, and a greater quality of life.

CHAPTER THREE

How Amino Acids Affect Your Muscles

When I was younger and trying to improve my chances of making the high school basketball team, I never thought of muscle beyond its importance in strength and physical function. My focus was purely on getting bigger and stronger.

Even today, that is the way most people think about muscle.

However, there is much more to muscle than just size and strength. Maintaining or improving our physical function by working on our muscles is a vital part of overall health … yours included.

Playing Catch Up

The first time we are likely to really think about physical function may be when we have trouble doing something that used to be routine, such as walking up a flight of stairs or carrying the groceries in from the car. It could be after surgery or, maybe, a prolonged period of drinking too much alcohol when we realize how much weaker we feel.

Unfortunately, once we have started to lose a significant amount of strength, our stability and mobility are negatively affected, and it is difficult to reverse the decline. It is like closing the barn door after the horses have gotten out. Playing catch up is always hard to do.

> The best time to begin using essential amino acids … is before you realize how much you need them!

The primary goal with essential amino acids is to get them into your body before the proverbial horses get out of the barn. Of course, starting late is better than not starting at all, but the optimal time is to begin before you realize how much you need it.

This is particularly true in regard to muscle, and it has nothing to do with big muscles. Rather, it has everything to do with being able to function physically as we want and need.

I once told my 95-year-old mother about the central role that muscle plays in the regulation of metabolism in the body and the support of bone health. It enables greater survival in the case of heart failure and cancer, even affording psychological support. I told her about my work with essential amino acids and how they would help her in her daily life by not only strengthening muscle, but also improving all of her other functions that relate to her muscles.

Anabolism: where protein synthesis exceeds the rate of protein breakdown. You are strengthened and gain muscle.

She politely listened, then responded, "What do I need muscle for? I am just playing bridge all day. I don't need to be strong for that."

My experiences with many, many more people since then is that most people only think of physical performance, such as size, bulk, and strength, when thinking about the importance of muscle.

That is so not the case.

Muscle Keeps You Balanced

Muscle has always been recognized for its importance in mobility and physical activity, but did you know that muscle plays a big part in maintaining the metabolism of protein in your body?

As we have already discussed, your body is composed of proteins that are in a constant state of breakdown and synthesis. Their goal is to stay balanced between anabolism (gaining protein) and catabolism (losing protein).

Throughout the day, your body routinely goes through periods of both anabolism and catabolism, depending on whether you have just eaten a meal or whether it has been several hours since you have eaten and you are no longer absorbing amino acids.

Thankfully, your organs and tissues maintain the balance between synthesis and breakdown throughout the day as well … even if you are not consuming dietary protein.

This is a good thing! Just think of the problems we would have if we missed a couple of meals and the skin protein became catabolic and caused us to lose a significant amount of skin. Or what if we lost protein from the liver, heart, or kidneys?

> Catabolism: where protein breakdown exceeds the rate of protein synthesis. You are weakened and lose muscle.

The essential tissues and organs maintain a balance between protein synthesis and breakdown in the absence of dietary protein consumption because they can draw from the amino acids circulating in the blood. Even in the absence of food intake and continuous uptake of amino acids from blood for protein synthesis in tissues other than muscle, the blood amino acid concentrations remain constant.

How does your body stay balanced? You can thank your muscles for that!

Muscle Keeps You Alive

The body is remarkably efficient in maintaining constant concentrations of amino acids in the blood. This is true even in the case of prolonged fasting.

In the early 1960s, it was common to place severely obese individuals on supervised starvation. One of the most famous cases involved an individual who weighed more than 700 pounds. This individual was provided only water, minerals, and vitamins for over a year, with the result being a loss of more than 500 pounds.

Despite going for over a year without protein (or any other) macronutrient intake, plasma amino acids were maintained at the same level as at the start of the diet.

Further proof of the vital importance of maintaining the circulating levels of amino acids comes from an unlikely source: the IRA hunger strikers who protested the British control of Northern Ireland. They chose to starve themselves to death to draw attention to their cause. They insisted that blood samples be taken throughout the process so some medical good could come of their sacrifices.

None of them were obese and, sadly, they died in about 40 days. The key signal that death was close … blood amino acid levels could no longer be maintained.

When the amino acid levels dropped below the normal level, the synthesis of essential proteins could not be sustained, and they died.

Muscle plays a key role in maintaining the plasma amino acids' levels in the absence of absorption of dietary amino acids from digested protein. You can consider muscle to be the reservoir of amino acids for the rest of the body.

Stress increases your body's demand for amino acids.

It is the only tissue in the body that can afford to lose some of its mass without impairment of health. In the absence of dietary amino acids, there is a net breakdown of muscle protein to supply amino acids to the blood to balance the amount taken up by the tissues in order to maintain health in other tissues and organs. The result is a net loss of skeletal muscle in the absence of dietary protein intake.

In short:

> Your muscles sacrifice themselves so that you can live.

That is precisely what muscle atrophy is. Your body is sacrificing its muscles in an effort to get the essential amino acids to your vital organs!

If you are not eating a balance of essential amino acids (the average person does not eat a balance of essential amino acids and a body after age 50 does not efficiently synthesize protein), atrophy is a common result.

Thankfully, the situation can be reversed if caught in time. Muscle loss can be replaced by getting a balanced formula of essential amino acids in foods and with the essential amino acid solution (EAAS) formula.

Eating dietary amino acids usually results in the production of muscle and, ideally, those essential amino acids are in the proper amounts and combinations to get the best results.

CHAPTER FOUR

The Surprising Benefits of Stronger Muscles

Muscle strength is not all about looking good, competition, or showing off. Muscles play a much bigger role than that. In fact, the benefits of your muscles might amaze you!

Muscle Helps Fight Diabetes

Muscle plays a key role in regulating the blood concentration of glucose (blood sugar), as well as the amino acids.

Under normal conditions, the brain relies entirely on glucose from the blood for energy. A drop in blood glucose concentration can cause loss of consciousness and even death. In contrast, an increase in glucose concentration in the blood is responsible for many of the adverse effects of diabetes.

It is important to understand:

> All dietary carbohydrates are ultimately converted to glucose in order to be metabolized in the body.

After you eat carbs, your blood glucose level increases. The hormone insulin works to moderate this increase by stimulating the uptake of glucose, mostly by the muscle. Once in the muscle cell, the glucose is converted to a chemical form of energy or stored as glycogen for later use.

Muscle is not only important in blunting the magnitude of increases in blood glucose after meals. It also helps prevent decreases in the blood glucose level between meals that could impair brain function. This is because the liver can produce new glucose from amino acids when you are not absorbing dietary carbohydrates.

> The loss of muscle strength is more strongly related to mortality than the amount of muscle mass.

What that means to you is this:

> Maintaining healthy muscles is crucial for keeping your blood glucose levels in the normal range.

Keeping your muscles healthy is vital for the prevention of diabetes, as well as other health problems caused by hypoglycemia (low blood glucose level).

Who knew muscles were so important, right? But that is not all.

Muscle Helps Fight Bone Loss

Muscle also helps fight bone loss. We know that weight-bearing exercises increase not only muscle strength but also bone strength.

Prevention of the loss of bone with aging (osteoporosis) is highly dependent on the maintenance of adequate muscle mass and function.

That is because strong muscles put tension on the bone, and that is vital for healthy bone function.

Muscle Helps Fight Cancer

Cancer is the most documented clinical state in which survival is directly linked to the maintenance of muscle mass. Cancer is associated with a rapid loss of muscle mass and strength at a rate faster than would normally occur because of decreased protein intake alone.

Survival from a variety of cancers is directly related to how well the muscle mass is maintained.

In addition, the ability to withstand the rigors of chemotherapy and radiation therapy is directly related to muscle mass and strength.

> Your muscles play an important role in regulating your blood glucose levels.

Muscle Helps Strengthen Your Heart

Heart attacks and strokes are more common and usually more disastrous in individuals with depleted muscle mass. To say it another way:

Those with healthier muscles live longer.

Survival from other serious diseases, such as chronic obstructive lung disease and heart failure, is also better in individuals with greater muscle mass.

Muscle Helps Fight Obesity

Muscle also plays a very important role in energy balance and prevention of obesity. The process of continuous synthesis and breakdown of muscle can be calculated. Here is the equation:

$$10 \text{ kg of lean muscle} \approx 100 \text{ kcal/d}$$

That means a person who has 30 kg more muscle mass than another person will burn approximately 300 kcal more energy every single day at rest ... and even more during exercise.

The results:

> The person with more muscle mass can burn off as much as 30 pounds of fat in a year, all without trying!

Muscle burns fat because muscle is always burning, even while you are sleeping. This is another powerful reason for working to increase your muscle mass.

So, whether you are fighting diabetes, bone loss, cancer, heart issues, obesity, or the gradual downhill slide of aging, the best way to defend yourself is with strong muscles.

Keep working on your muscles, regardless of your age. You will be so glad you did!

CHAPTER FIVE

What Is Healthy, Anyway?

Most people want to "eat healthy," "live healthy," and enjoy all the benefits of "healthy living," but how do you know what "healthy" really is?

Because we are inundated with advertisements and scientific "breakthroughs" that recommend an incredible range of foods, supplements, activities, and exercises that are all supposed to offer great health benefits, just what is "healthy" is very hard to define.

Where do you begin? And when you do start, how do you know you are going in the right direction?

How Much Protein Is Healthy?

One good starting point is protein, the only dietary component we absolutely have to eat in order to survive. Without it, we would die.

The government's Recommended Dietary Allowance (RDA) for protein says you need to eat at least 0.80g of protein/kg body weight/day. For a 175-pound person, this translates to:

About 65 g or 2.2 oz of protein per day.

Now, an average American diet consists of 4-5 oz of protein per day, which is about twice the RDA. There are many sources of protein in the diet, such as eggs, yogurt, cheese, milk, nuts, ice cream, pudding, fish, chicken, steak, wheat, peas, beans, and soy.

Since the RDA is the minimal amount, eating more than 2.2 oz of protein a day is certainly within their range of reasonable consumption.

But what about essential amino acids? Is there a RDA for essential amino acids?

> Your body probably needs more essential amino acids and protein than you are giving it.

Since they cannot be produced in the body and must be obtained from the food we eat, it would make sense that we follow the RDA for essential amino acids.

Unfortunately, there have been very few studies to define the RDA for each individual essential amino acid, and those studies that have attempted to do so are deeply flawed.

As a result, the RDAs for the essential amino acids are set very low, so low in fact that even a diet considered deficient in protein would still satisfy the RDAs for the essential amino acids.

To make matters worse, essential amino acids must be balanced, and if they are not, the RDA itself is of little value.

I have found, after years of study and research, that:

> Optimal essential amino acid and protein nutrition translates to a significantly higher intake of high-quality protein and essential amino acids than the RDA states.

Simply stated, if you have been trying to get your recommended daily allowances of proteins, your body probably needs more protein and more essential amino acids than you are currently getting. There is no danger that you are eating too many essential amino acids.

How Can I Keep My Brain Healthy?

As the population ages, many people are becoming aware of the importance of "exercising" the brain and feeding it the proper nutrients or supplements to keep it sharp.

Without question, the brain is the most important organ in the body. It is so central to our being and life that it is hard to think of it as just another tissue that requires oxygen and nutrients to keep functioning optimally.

Because the brain is so vital to life, it is well protected by a specialized structure that only lets certain molecules pass through into the brain. This protective shield is called the blood brain barrier.

> Essential amino acids serve as important chemical messengers in the brain.

Across this protective shield flow the essential amino acids, which become important chemical messengers that can affect mood, appetite, energy level, sex drive, and many other behaviors and feelings that affect our lives.

With sufficient amounts of the right essential amino acids, the brain is able to function at its optimal level.

How Can I Keep My Liver Healthy?

The liver detoxifies and purifies blood, but when the liver gets fatty, your body pays the price.

Specifically, your liver plays an important role in amino acid metabolism. Amino acids from protein digestion get sorted and transformed into different (non-essential) amino acids in the liver, depending on the need. The liver helps to maintain a proper balance of amino acid concentrations in the blood by producing non-essential amino acids that might be in low supply.

Drinking too much alcohol or being overweight will cause fatty liver, but unfortunately:

> The normal process of aging is associated with increased liver fat.

As the liver cleans your blood, normally only a small amount of fatty acids from the blood are stored in the liver. A healthy liver repackages these fatty acids and secretes them back into the blood to be delivered elsewhere in the body for storage.

But if the liver itself begins to store fat, it is a sign of metabolic dysfunction. A fatty liver leads to diabetes, hepatitis, scarring of the liver tissue, and serious liver diseases, including cirrhosis.

Traditional medicine, however, is not very effective at treating a fatty liver. Commonly prescribed medicines often have adverse side effects, especially in older individuals.

> Essential amino acids help keep the liver functioning optimally.

I have found that regular consumption of the EAAS formula is effective in treating a fatty liver … and with no adverse side effects!

Not only does your muscle play a part in maintaining relatively constant levels of amino acids in the blood, so too does your liver. It is for this reason that all "healthy" dietary recommendations should include a balanced approach to essential amino acids and protein nutrition.

Your overall health depends on it.

CHAPTER SIX

Muscle Loss with Aging

Scientists argue about many things, but one point not debated is the fact we lose muscle mass and strength as we get older.

Nobody has ever come up to me and said, "That is not true; I'm getting stronger the older I get."

This fact applies to everyone:

> We all lose muscle as we age.

For about 30% of people, the loss of muscle becomes severe. This is called sarcopenia. Once someone suffers from sarcopenia, the functions of daily living are severely affected.

But the fact is you do not need to be diagnosed with sarcopenia to suffer adverse health consequences as a result of even a modest amount of muscle loss.

Consequences of Muscle Loss

Muscle loss has many negative ramifications. It is well recognized that you cannot run as fast or hit a golf ball as far as you get older.

For most people, they do not expect to run as fast or play a sport as well. They are fine with that, but what they often do not realize is:

> The gradual loss of muscle strength affects daily living and can eventually hurt their quality of life.

Loss of muscle mass and strength has much broader health implications. Recent research has made clear that significant loss of muscle mass and/or strength will:

> Muscle loss has a way of always negatively affecting your life.

- increase your risk of cardiovascular events
- decrease survival from various diseases, including cancer and chronic obstructive lung disease
- impair recovery from major surgery
- cause bone health to suffer

These health issues are all related to muscle mass. The reason that muscle is at the core of all these health conditions is due to muscle's role as the reservoir for amino acids.

Amino acids from your muscles are mobilized when other tissues and organs need an increased supply of amino acids, such as to battle infection, repair wounds, gain vascular control, achieve metabolic balance, and more.

Muscle Loss Begins Before You Realize It

Most often, people do not recognize they have lost muscle mass and function until they are 70 years old or older.

This oversight is particularly the case with people who do not participate in organized recreational activities. Because they do not get feedback on their performance, they do not recognize the problems caused by the loss of muscle mass and function.

The loss of muscle can start in some people as early as age 30, but by age 50, almost everyone is starting to lose a significant amount of muscle.

It is typically hard to notice muscle loss for two main reasons:

#1—Your body weight does not change or you might actually be gaining weight

#2—You can still comfortably perform all the activities of normal daily living

As the loss of muscle progresses, basic function may still be maintained. Usually, this will be the case until there is some health set back. Then suddenly (from serious illness, injury, or surgery) there is a noticeable amount of muscle loss that starts to affect the performance of simple activities.

At that point, people ask, "What happened? How did that happen so fast? What's going on with my body?"

What happened is that the muscle loss suddenly became apparent. That is all.

Muscle also plays a role in maintaining a healthy energy balance.

When the normal, age-related rate of muscle loss is coupled with the accelerated loss that occurs in response to a health crisis, your body suddenly seems to "age."

Reverse and Prevent Muscle Loss

Thankfully, it is possible to reverse muscle loss and to prevent muscle loss. That is the beauty of the right balance and ratio of essential amino acids!

With a steady intake of the essential amino acid solution (EAAS) formula, along with an exercise program, your body can maintain mass and even repair itself from muscle loss.

As we age, it is much easier to maintain muscle mass than to regain it once lost. That holds true for two primary reasons:

> #1—After you have lost a significant amount of function, you are limited in the

Why Aging Affects Muscles

As you age, the rate of muscle synthesis slows down, and the rate of muscle breakdown speeds up.

Thus, normal protein nutrition is not as effective in the elderly as in younger people. That is the main reason you lose muscle as you age.

amount of exercise you are able to perform.

#2—Once muscle is depleted, there are metabolic changes that make it less receptive to the beneficial effects of essential amino acids.

The bottom line is to try to maintain your muscle mass and function before you lose it. When you turn 50, it is time to get serious about maintaining your muscle.

The right balance and ratio of essential amino acids are of paramount importance at this point.

CHAPTER SEVEN

Handling Hormones As We Get Older

Hormones change as we age. This affects all of us, men and women included.

Much of these hormonal changes are a natural part of aging, but when those changes affect your body and muscles, it can really make an impact in your quality of life.

Hormones and Women

Middle age in women is marked in part by menopause, which occurs around the age of 50. Many changes occur with menopause, including the end of menstruation.

The most important aspect of menopause is probably the reduction in secretion of female hormones from the ovaries. Most notably, estrogen secretion is reduced, but the secretion of other hormones is also reduced.

Decreased secretion of estrogen after menopause contributes primarily to the deterioration in bone health. Hormone replacement therapy (HRT) with estrogen and progestin has been used to counter the symptoms of menopause.

The use of HRT has declined in recent years due to the side effects and complications (like blood clotting and strokes). Other medications are now often prescribed for a more targeted therapy of specific symptoms of menopause.

> Hormone treatments often include adverse side effects.

Hormones and Men

The predominant hormonal response to aging in men is reduced secretion of the hormone testosterone from the testes.

Testosterone promotes the gain of new muscle protein and has other effects as well. Sexual function in older men is directly affected by this reduction in testosterone.

Replacement therapy with testosterone is popular, but therapy is limited by the fact that testosterone cannot be given as a pill. Patches are commonly used to increase testosterone levels, but the amount of the hormone that can be delivered by this route is limited and insufficient to significantly affect muscle. To benefit from testosterone therapy:

> Testosterone must be injected (usually once a week or once every two weeks) to increase the concentration of testosterone enough to have an effect on muscle.

Only some aspects of testosterone action can be restored by the use of testosterone patches. Most notably, sexual function, which includes both the level of interest in sex as well as the ability to do something about that interest, is improved when testosterone is given in a patch format.

Like hormone replacement therapy, they also have testosterone replacement therapy for older men. However, it is controversial because of worry that the growth of any existent prostate cancer may be stimulated by testosterone therapy. The first line of action in the treatment of prostate cancer is giving a drug that blocks the action of testosterone on the prostate.

As with women, replacement therapy is used cautiously.

Other Hormones Change with Age

There are other hormones that affect both men and women equally. This holds true with age.

For example, insulin plays a crucial role in regulating blood glucose levels. The development of diabetes is directly related to the loss of normal insulin action. Insulin stimulates muscle protein synthesis and inhibits muscle protein breakdown, the exact opposite of the tell-tale sign of aging.

As you age, the function of insulin decreases, noticeably in most people over age 65. This affects the way the body handles carbohydrates and protein, and may also contribute to loss of muscle mass.

Another hormone that is greatly reduced as you age is your growth hormone. This is to be expected. However, in an effort to slow down this hormone loss, growth hormone therapy was introduced.

Growth hormone therapy also has some negative side effects (such as increased insulin resistance), but this is still popular with older people.

Sadly, there is no evidence of a beneficial effect of growth hormone therapy. This is because our bodies lose responsiveness to growth hormones as we age. After all, we stopped growing many years ago!

How to Help Your Hormones

Changes in hormonal secretion are a natural part of aging. No doubt you are thinking:

> "Okay, so what can I do, without negative side effects, to help my body with its hormone levels?"

The essential amino acid solution (EAAS) formula with its ratios of the nine EAAs has no negative side effects and does not require a physician's supervision. However, the formula is not a specific, targeted hormone therapy treatment plan in and of itself.

Helpful for anyone at any age, the essential amino acid solution (EAAS) formula plays a proven and effective role in your muscles.

Consider testosterone treatment. Testosterone is similar to resistance exercise in that it primes the muscle to increase its rate of synthesis. Remember, it needs building blocks (essential amino acids) to actually produce new protein. The essential amino acid solution (EAAS) formula will amplify any beneficial effects of testosterone on muscle protein synthesis.

> Older people have, on average, twice the fat in their livers than when they were under 30 years of age.

Or consider insulin resistance. One of the characteristics of insulin resistance with aging is that fat accumulates in the liver and further limits insulin action as a result. The essential amino acid solution (EAAS) formula reduces liver fat without the adverse effects of medicine, and insulin sensitivity is improved. Exercise also improves insulin sensitivity.

The fact that the EAAS formula so positively affects muscles and overall body health is another reason why it is an effective way to stay vibrant and healthy as you age.

CHAPTER EIGHT

The Right Formula for Living Stronger

Ironically, the usual intake of essential amino acids tends to decrease in older people just when increasing it would really make a positive impact on their lives. In America, 30% of people over the age of 65 fail to eat the minimal daily recommended intake of protein, which is the conventional dietary source of essential amino acids.

The consequences of the lack of essential amino acids in the proper ratio and balance become functionally evident with advancing age. As we have discussed, not just any ratio of an essential amino acid supplement will do.

Intake needs to be balanced and in the right ratio, such as the EAAS formula, if you want to curb the consequences before they are established.

No matter where you are, it is comforting to know this fact:

> It is never too late to do something about loss of muscle mass and function.

Trying to Meet a Need

Furthermore, the quality of the dietary protein also decreases with many older people, which means that essential amino acid intake is greatly reduced.

Most prominently, meat intake decreases. This may occur for a variety of reasons, including high cost for people living on a fixed income, issues related to food preparation, problems chewing, digestive limitations, and changes in taste preferences.

Unfortunately, people may also be reacting to advice to decrease their meat intake to benefit

> Loss of muscle becomes significant by age 50. Now is always the time to get serious about slowing the progression, if you have not started already.

their health. Any or all of these factors may lead to an inadequate intake of high-quality protein in meat.

Enter the dietary supplements, mostly with protein-enhanced beverages, that have been marketed as therapy with people over 65 years of age. The general idea is to provide some high-quality protein (like whey protein) to increase the essential amino acid content of the diet since the typical diet is lacking in protein.

> Hormonal changes with aging contribute to the loss of muscle.

However, these types of protein supplements have not proven to be consistently helpful in older people. The problem is the same as why we lose muscle as we age:

> The normal ability of dietary protein to stimulate muscle protein synthesis is diminished as we age.

Consequently, even if enough supplemental protein is provided to bring the total essential amino acid intake level up to an optimal level, loss of muscle mass and function will not be reversed.

Finding the Right Formula

One major advantage of the essential amino acid solution (EAAS) formula is that it has been precisely adjusted to specifically stimulate muscle protein synthesis in the aging population.

After years of laboratory research, we were able to determine the optimal profile of essential amino acids to maximumly stimulate muscle protein synthesis.

We found our ideal mixture of essential amino acids equally brought about improved muscle protein synthesis in younger people and older people.

That was quite a breakthrough considering the fact that stimulating muscle protein synthesis in older individuals is an age-old challenge.

The difference between the effectiveness of essential amino acids in our EAAS formula and

> The right essential amino acid formula gives you what your body needs, which would not have been possible without eating a large and perhaps stringent amount of foods.

regular protein supplements (like whey protein) was so great that there is no comparison.

It is a matter of quality, not quantity. The right essential amino acid formula gives you what your body needs, which would not have been possible without eating a large and perhaps stringent amount of foods.

We even successfully used this formulation of essential amino acids to decrease the loss of muscle mass and strength that occurs with bed rest and recovery from hip replacement.

The essential amino acid solution (EAAS) formula proved to be exactly what our bodies need as we age.

CHAPTER NINE

Helping Your Body to Heal

At some point in your life, you may have to deal with major surgery or serious illness. There may be a phase in your treatment when nutritional therapy is part of your care, but more likely than not you will be on your own for taking care of your nutritional needs.

This is particularly true for recovery from orthopedic surgery that many of us need as we age. It is also the case in recovery from any type of serious illness, ranging from a bad case of the flu to something so serious that you end up in intensive care.

Even with the recognized importance of the role of nutrition in cancer recovery, detailed nutritional advice is rarely provided in standard treatment programs. I worked for decades on the nutritional and metabolic care of patients with severe burns and other forms of critical illness. The essential amino acid solution (EAAS) formula helped them tremendously.

So rest assured that what you are reading is based on solid nutritional and metabolic principles. It is also consistent with nutritional guidelines provided by all national and international committees involved with the care of seriously ill patients.

The Big Stress Response

Muscle loss as a result of serious injury or illness is far more severe than the result of aging. With aging, the loss of muscle mass is slow and occurs over many years. A serious illness or injury, on the other hand, causes muscle loss in a matter of days or weeks.

> Many metabolic problems accompany critical illness.

Because the rate of muscle loss with aging is so slow, it is difficult to determine the underlying cause for the loss. But with an injury or serious illness, the exact cause is usually known.

This rapid loss of muscle is part of an overall response that can be considered a "physiological stress response." There are two primary and common aspects of a stress response, no matter what the underlying cause (i.e. cancer, injury, surgery) might have been that directly affect your muscles. These are the following:

> #1—Muscle loss, which is much faster than usually occurs from not eating food.

> #2—Loss of appetite, which leads to further decreased nutritional intake.

The rapid loss of muscle protein reflects an imbalance between the rates of protein synthesis and breakdown. It is like fast-forward aging!

This imbalance between protein synthesis and breakdown is caused by a massive increase in the rate of muscle protein breakdown, sometimes 3x the usual rate. The body does try to balance this by stimulating muscle protein synthesis, but it is not enough.

The stress response also kills the appetite so much that it is hard to eat any food. Even when you do force yourself to eat or rely on meal replacement beverages, you see little or no beneficial response.

The end result of these two debilitating factors is muscle loss, and usually on a large scale.

How to Bounce Back

The catastrophic results of a stress response are an excellent example of why the balance between the rates of synthesis and breakdown is so important.

But after a stress response situation where rapid muscle loss has occurred, is it possible for the body to bounce back?

The answer is "yes," but the body needs a little help.

> Muscle mass in critical illness is a direct contributor to survival, as well as to the speed and extent of recovery.

Eating the right foods and proteins will help, but the resulting shift back toward balance will take a long time … without the aid of the essential amino acid solution (EAAS) formula.

One result of the stress response and severe muscle loss is that the body is unable to process essential amino acids like it usually can. It is like filling up a glass with water, but when you turn around to drink it, all but a tablespoon has leaked out.

A meal or protein beverage is unable to give your body the essential amino acids it needs after a severe stress response. What should you do?

> You need to flood your body with the nine essential amino acids of the EAAS formula, which contains the right balance and ratio.

That is the only way to counter balance the extreme muscle loss. By doing so, you will reactivate the molecules that are involved in initiating protein synthesis.

By consuming the essential amino acid solution (EAAS) formula, which provides more essential amino acids than you can get from food alone, your blood and muscle concentrations will reach high

Exercise is always important in relation to muscle mass and function, and never more so than in recovery from illness, injury, or surgery.

enough levels to stimulate muscle protein synthesis.

In addition, if enough of the EAAS formula is consumed, the concentrations will rise high enough to stop the muscle protein breakdown. As a direct result, your body is able to bounce back!

Other Effects of Stress

In a stress response situation, there is more that happens to your body than the accelerated rate of muscle protein breakdown causing muscle loss. Even though that is more than enough and tough to deal with, sadly it is not all.

Here are several hallmark signs of stress response in your body:

- Stress causes increased insulin resistance, which prevents the liver from registering the message to stop producing glucose. As a result, blood glucose levels rise uncontrollably.

- Stress causes carbohydrates to be stored as fat rather than used for energy production. Some of that fat is then stored in the liver.

- Stress causes adrenaline, one of the main stress hormones, to pump fatty acids into the bloodstream. The result is that there is an excessive amount of circulating fatty acids, more than is needed or can be used for energy. This again affects the body and the liver.

When we are faced with intense physiological stress and our bodies cannot handle the increased demand for energy, our bodies literally pay the price. Sometimes it is fatal, but more often than not, our bodies will slowly recover.

It is at this moment that the essential amino acid solution (EAAS) formula is so vital. Perhaps the stress response could have been avoided entirely with sufficient essential amino acids on board, but as a remedy, nothing is better than the essential amino acid solution (EAAS) formula.

CHAPTER TEN

Nothing Can Keep You Down

Illness, injury, and surgery bring about inactivity. That is completely normal.

If you are ever in an intensive care unit, being confined to bed is the norm. In fact, if you are over age 65, you may be confined to bed in the hospital even when you are fully capable of walking. They are concerned about your stability and the risk of a fall, so they confine you to bed.

But here is a fact that hurts:

> Being confined to a bed is not good for your body.

After a surgery, you are likely to be physically limited until incisions are healed and you have some degree of recovery.

In all cases of serious illness, you probably have no desire to exercise. Your body hurts, so rest and recovery are far more pressing needs than exercise.

That also is normal, but inactivity has a way of multiplying the negative effects of a stress response, illness, or hospital visit.

The Multiplication of Inactivity

Scientists often wonder, "How do we know the degree to which inactivity amplifies the problems that occur due to physiological stress?"

You may have wondered the same thing a time or two, but we found part of the answer to that question by a series of studies sponsored by NASA. Like being confined to bed, the lack of gravity during spaceflight reduces the physical work of movement drastically, with the consequence that muscle loss during space flight can be severe.

> Inactivity amplifies the loss of muscle mass in the stress response.

Odd as it may seem, enforced bed rest has been used by NASA for many years as a model for the effects of the lack of gravity during space flight. We performed a number of these studies with research volunteers who maintained complete bed rest for up to a month. (Studies have not only been done in young, healthy subjects, but in older subjects as well.)

As you might expect, inactivity causes the opposite effects of exercise. Muscle mass and strength are reduced in bed rest. Bed rest also induces insulin resistance. (The rate of deterioration of these factors is faster in older, as opposed to young subjects confined to bed rest.)

The results were always the same:

> Inactivity in and of itself is detrimental to muscle mass and function. When coupled with stress, bed rest amplifies the muscle loss.

To try to better understand the role of inactivity, we did a study with healthy individuals who completed a bed rest trial. We then administered cortisol, a hormone that is considered one of the mediators of the stress response. We compared the response to cortisol when subjects were up and active to their response to cortisol when they were in bed rest.

Both bed rest and cortisol independently decrease muscle mass, strength, and function ... but the loss of muscle is much more rapid when the two factors are combined.

These studies taught us a great deal about metabolic changes brought about by serious illness or trauma. While the extent of physiological stress is less than in actual disease states such as cancer or severe injury, all of the same metabolic responses are evident.

Reversing the Effects of Inactivity

Next, being practical scientists, we were interested in what treatments could help minimize the loss of muscle and metabolic problems that occur in these conditions.

We discovered two answers:

> #1—Those who performed a minimal amount of resistance exercise preserved their muscle a lot longer than those who did not.

> #2—Those who took the essential amino acid solution (EAAS) formula mixture, multiple single servings per day, had the

best results with slowest loss of muscle mass and strength. This was true for young and old subjects alike.

We also tested these treatments in a clinical trial in which essential amino acids were shown to speed recovery from hip or knee replacement as well.

> The challenge in the recovery stage is to regain lost muscle mass without a significant gain in fat mass.

If you have lost significant muscle mass due to illness, injury, or surgery, you need to get your muscle back. You also need to make sure you do not replace the lost muscle with fat, as often happens due to inactivity and lack of essential amino acids.

Multiple single servings daily of the essential amino acid solution (EAAS) formula are proven to improve muscle mass and strength considerably. They are also the necessary ingredients for your recovery.

In the recovery phase, your body is healing and fighting against the metabolic responses. Also, in recovery, you do not feel like exercising or you may not physically be able to.

The answer is a balanced ratio of the nine essential amino acids found in the essential amino acid solution (EAAS) formula. This provides a unique stimulus to muscle protein synthesis, which in turn speeds up your recovery.

As you age, the only thing you want to speed up is your recovery time.

CHAPTER ELEVEN

The Reality of Exercise

I have found that most people reading this book will fall into one of two categories. You may as well:

> #1—You have not done much physical activity for years.

> OR

> #2—You still work out and/or participate in active sports, such as golf, tennis, or jogging.

I have also found that when it comes to exercise, we need to be very real. The reality is that we all need to exercise. The other reality is that some people are not able or do not want to do much exercise.

How you handle that reality, whatever your reality might be, will play a large part in your ongoing strength, mobility, and quality of life.

Exercise Plan #1—Starting Up

If you have not done any kind of exercise for years, you may benefit from an exercise trainer to develop a program specifically designed for your capabilities and goals.

Regardless of the program, you will have to learn how to push yourself again, but you are in the enviable position where you will see rapid improvement as you begin to train, whatever the exercise you choose.

The improvement will be especially dramatic if you couple your exercise with the essential amino acid solution (EAAS) formula. Your improvement will inspire you to stay with whatever program you have laid out for yourself.

Be aware that at some point you will plateau in your progress as your body catches up to your exercise regimen. This is natural and means it is time to push harder to keep improving.

> Exercise is the best way to reverse muscle loss and regain normal function.

The key is consistency. You have to make your exercise a non-negotiable part of your day. So go ahead and set a rigid schedule and adhere to it. That is the surest way to get the results you want.

Exercise Plan #2—Continuing Ahead

If you have exercised all your life, your perspective will be quite different than if you are starting up an exercise program. Although you have the tremendous advantage that you are probably in much better shape than the average person your age, you still face challenges. The challenges are part physical and part psychological, and the two aspects may merge together indistinguishably.

I can relate some of my experiences in this regard, as I fall into this group and have talked with many former athletes in the same situation.

When I was young, my athletic focus was basketball. After that phase was finished, I took up distance running seriously. Although not a world-class runner, I nonetheless embraced the

> You may initially need a physical therapist, followed by an experienced trainer, to formulate the best plan for recovery exercises.

challenge of setting both short-term and long-term goals and training hard to try to reach those goals.

I trained and raced steadily from the time I was 25 years old until age 69, when I had my hip replaced. Nine months after that surgery, I was successfully jogging again.

Being a scientist, I methodically approached training for marathons. I carefully recorded every workout in detail, with the distance, time, etc. I did interval workouts twice per week and recorded every split of every interval.

As I got older, I faced challenges that all of my friends who ran also faced. Of course, there were the usual injuries. They not only became more frequent as I got older, but it took longer to recover. But most discouragingly, I inevitably got progressively slower. I was 65 when I last trained for a full marathon. I was still trying to do the same program as when I was younger, but with drastically less success. I was running quarter-mile intervals at a considerably slower pace than I could maintain for an entire 26-mile marathon when I was younger.

Intellectually, I understood that the days of 65-second quarter-mile intervals were long in the rear view mirror, but, psychologically, it was much harder to accept. I would look at the workouts in my running log that I used to do and feel

My Hip Surgery

After my hip replacement, I was given a set of exercises that would seem pretty lame to someone who didn't just have their hip replaced. I was told to do the exercises 5 times a day, which was quite time consuming, but since I had nothing else to do, I did the simple exercises.

After a few weeks, I began to feel like I would never get back to where I wanted to be, but I knew from experience that there was no substitute for persistence. If I persevered, things would eventually get better … and they did.

My energy and strength did come back, as they will for you if you fully commit to the process.

discouraged at the decline in my ability. I could understand why almost everyone I knew from years of running had given it up.

At some point, I assessed things and decided that I would rather keep running at a slower pace and within my realistic capabilities than to quit it altogether. So I started a new running log, put the old ones away, and forgot about what I used to do.

I wish I could say that I saw the light and loved running again, but it did not really work that way. In fact, I still get frustrated when jogging and someone breezes past me like I'm standing still! Nonetheless, I have made enough peace with my current abilities that I keep plugging away every day. I now take naps after good workouts, as I do not recover very fast anymore.

The exercise is just what I do. I do not question whether I will do it or not, I simply choose to do it regardless of how I feel.

The lesson from my personal story is that the important thing as you get older is to keep doing whatever activity or sport that you like to do. You have to get past the dismay that you need to play from the forward tees on the golf course when you used to play from the tips, or that you can only play doubles in tennis now.

The benefits of participation in activities you enjoy will trump any of the negative aspects. The social aspect of joining workout classes in the gym or a group to play golf with regularly is an important step in the adjustment to aging or retirement.

> It is inevitable that you will SEE, FEEL, and ENJOY progress if you stick with your exercise program.

You just need to let go of what you used to do and be happy with what you can do now.

What Progress Looks Like

After about 4-6 weeks of committing to the basic diet and exercise recommendations, you will start to feel the effects. You will have more energy for the activities you enjoy, and your recovery time will be much faster.

You may not necessarily lose weight (muscle weighs more than fat) during those first 4-6 weeks, but your clothes will start fitting better. You will feel more fit.

The benefits and good feelings will snowball forward! Trust me.

That is what feeling younger is all about!

CHAPTER TWELVE

3 Easy Steps to Staying Younger

While it is impossible to entirely stop the process of aging, it is definitely possible to slow the progression of muscle loss.

That alone is a huge breakthrough!

There is no fountain of youth, but there are 3 steps to staying younger at any age:

1. A basic, healthy diet
2. The EAAS formula
3. Regular exercise

Combined, these 3 steps are incredibly effective at slowing down the rate of decline in muscle strength and mass that comes with aging.

Step #1—A Basic, Healthy Diet

Regardless of your age, a basic healthy diet applies to you. The most important element is to eat at least 25% of your calories in the form of high-quality protein.

Caloric equivalents of high-quality protein food sources:
Caloric Equivalents Calculated from Data in USDA Nutrient Data Base: (https://ndb.nal.usda.gov)

Protein food source (kcal/oz food source)	Energy as protein (kcal/oz food source)	Total energy as Protein	Percent Energy
Beef (90% lean ground beef)	28	51	57
Egg (whole, poached)	14	40	35
Milk (1% milk fat)	5	13	38
Yogurt (low fat)	18	29	62
Lamb (composite)	28	40	70
Deer	34	42	81

Caloric equivalents of plant-based protein food sources:

Soy Beans	13	41	32
Kidney Beans	6	23	25
Chickpeas	8	39	20
Mixed nuts	23	170	13
Seeds (Sunflower)	23	165	14

This list of high-quality protein food sources will help you figure out what this recommendation translates to in terms of protein food sources.

The remainder of the diet is flexible and should include a variety of fruits and vegetables, as well as whatever special treats you might enjoy.

The principles of the basic diet apply regardless of age, with one little bonus for older people:

> As you get older, your ideal body weight (as defined by body mass index) increases.

It is inevitable that you will pass through some health challenges as you age, and having a little extra weight gives you a cushion that can help in these circumstances. Consequently, you don't need to worry as much about putting on a few pounds as you did when you were younger. (Note: This may be the best thing about getting old!)

It is much better to be sure you eat enough than to think you are doing something good for yourself by eating lightly to lose some weight.

Of course, you cannot take this to an extreme, as obesity is the major factor limiting mobility in older people.

The protein component of a typical day of meals may look like this:

Example of dietary meal plan:

Breakfast:

2 eggs	14 kcal/oz x 1.65 oz/egg x 2 eggs	=	46 kcal
Yogurt	18 kcal/oz x 4 oz	=	72 kcal
		Total Breakfast:	**118 kcal**

Lunch:

Ham (6 oz)	33 kcal/oz x 6 oz	=	198 kcal
1 cup milk	5 kcal/oz x 8 oz	=	40 kcal
		Total Lunch:	**238 kcal**

Dinner:

Beef patty (8 oz)	28 kcal/oz x 8 oz	=	224 kcal
½ cup milk	5 kcal/oz x 4 oz	=	20 kcal
		Total Dinner:	**244 kcal**

Breakfast + Lunch + Dinner = 600 protein kcal

You will see that to eat 25% of your calories as high-quality protein can be a chore for many people. This is where supplementing your diet with the essential amino acid solution (EAAS) formula comes in.

Step #2—The EAAS Formula

The essential amino acid solution (EAAS) formula can be of benefit to you no matter your age, but it is a particularly important part of a diet in older people.

With aging, a number of factors can conspire to cause a decreased consumption of high-quality proteins. As a result, the amount of essential amino acids consumed is also likely to be reduced. This is not good.

About 25% of your diet should be high-quality protein.

Even if you do eat 25% of caloric intake as high-quality protein, as suggested here, your essential amino acid intake is probably below the level it was when you were younger.

This does not mean you should boost your dietary protein intake to the level of a younger person. Dramatic changes in your regular dietary pattern are difficult to stick with and can be stressful because of the time, effort, and potential cost involved in changing your routine.

Also, due to your increased age, your body cannot utilize the dietary proteins like it could when you were younger.

The most beneficial answer is to incorporate the essential amino acid solution (EAAS) formula into your overall eating habits.

As you know, the EAAS formula:

- Is much more effective than natural protein food sources in terms of stimulating muscle protein synthesis.
- Places less of a burden on the liver and kidneys than food proteins do.
- Causes an increased recycling of non-essential amino acids back into protein.

For best results, take your single serving (3.6 g) each day. This dosage will support your protein synthesis and, over time, gain in muscle mass. Those who wish to accelerate the positive effects can take the essential amino acid solution (EAAS) formula twice per day between meals.

Smaller and less frequent doses are also effective in stimulating muscle protein synthesis, but the effectiveness is reduced in proportion to the reduction in dose. A dose as small as 3 g has the equivalent effect of as much as 20 g of high-quality protein on stimulating muscle protein synthesis.

The EAAS formula serving size: 3.6 g between meals.

The dose of essential amino acid solution (EAAS) formula you choose to use will depend on a number of factors. If you are eating a diet relatively deficient in high-quality protein, then larger doses of the EAAS formula are essential. If you are exercising, the timing of the EAAS formula should be coordinated with the exercise.

Step #3—Regular Exercise

Exercise plays a pivotal role in maintaining muscle mass and function. It is, many would argue, the key to a happy and productive retirement.

The essential amino acid solution (EAAS) formula plays a unique role when it comes to exercise:

> The EAAS formula amplifies the beneficial effects of exercise.

Exercise alone has tremendous benefits, but the addition of the EAAS formula to the mix takes everything to a whole new level!

The most effective way to maintain muscle mass and strength is through resistance exercise. And no, you are not too old to do resistance exercise. I have seen people in their 90s improve their muscle strength and function as a result of resistance exercise.

Detailed instructions are not necessary, as studies have shown that the exact nature of the exercises make much less difference than the total amount of weight lifted.

For example, 3 sets of 10 repetitions lifting 20 lbs will give you the same result as 3 sets of 4 repetitions lifting 50 lbs. Multiple repetitions with lighter weight is less likely to result in injury.

Here is the fact of the matter:

> It doesn't really matter exactly what exercises you do, as long as you exercise a variety of muscles and lift enough weight to feel the strain.

The important thing to remember about resistance exercise is that you are priming your muscle to respond to the essential amino acid solution (EAAS) formula. That is because resistance exercise cannot increase muscle protein synthesis in a significant way without the essential amino acids from which to make the protein.

> You need regular exercise (preferably resistance exercise) twice per week and aerobic exercise four days a week. Do activities you like!

Practically speaking, if you take the EAAS formula 30 minutes before working out, you prevent the breakdown of muscle protein that would otherwise occur during exercise.

Taking another dose after the workout will amplify the exercise effect on muscle protein synthesis, with the net being a greater increase in both mass and function than would occur in response to either exercise or the EAAS formula alone.

There is also a beneficial interaction between the essential amino acid solution (EAAS) formula and aerobic exercise. In the case of aerobic exercise, the EAAS formula should be taken after exercise, since a significant portion of the essential amino acids taken before exercise will be oxidized during the exercise.

My recommendations for exercise are intentionally flexible. It is up to you. The only absolute is that you take the EAAS formula before and after resistance exercise and after aerobic exercise.

What you choose to do for your exercise program is up to you. Just make sure you do it!

Conclusion

A basic, healthy diet and physical exercise, coupled with the balanced ratio of the 9 essential amino acids found in the EAAS formula, makes for an explosive combination!

Your body will feel the tremendous benefits of increased vitality and muscle strength, all without negative side effects.

This can literally happen within a few short weeks!

Most importantly and fundamentally, as this becomes a way of life, you will be on your way to achieving long-lasting health. With good health, you have the quality of life that you both want and need.

My hope is that you will try the essential amino acid solution (EAAS) formula. I am sure you will enjoy the benefits!

—Robert R. Wolfe, PhD

LIVE STRONGER AT ANY AGE!

WHAT THEY'RE SAYING ABOUT MYOHEALTH:

"*I've always been motivated and in pretty good physical shape, starting as a wrestler in 6th grade and progressing into Jiu Jitsu as an adult. But after I became a dad raising a 4-year-old, my mind was focused on financial matters. I currently work two jobs—one as a firefighter—and also own a small business, and finding time to hit the gym can seem impossible.*

After I discovered MyoHealth powder with the EAAs and also the Vegan protein mix, I've regained my energy and been hitting the gym 6 days a week. I feel myself getting stronger, I'm happy to be a client of TriVita and can't wait till I achieve my optimum physique by continuing to eat right, exercise well and recover with great nutrients by TriVita! "

— Abraham C.

"*I'm 64, slightly overweight and have had 2 knee operations. My job includes looking after a fit and energetic young autistic man and we often hike together for exercise. While it normally takes us 90 minutes and leaves me huffing and puffing and my muscles aching, since taking MyoHealth we now complete our hike in 75 minutes and I leave my young companion out of breath.*"

— Bernard O. ★

"*Since I started taking MyoHealth and completed the 30-Day Challenge, I've experienced increased energy and the muscles everywhere on my body have begun taking on new dimensions. My physiotherapist of 10 years asked why I suddenly have the calf muscles of a cyclist. Little does she know that I haven't been on a bike in 20 years.*

MyoHealth is truly an amazing product!"

— Giles G. ★

★ ITBOs (Independent TriVita Business Owners) may have received remuneration for products sold.

ARE YOU UP FOR A CHALLENGE?

The MyoHealth™ 30-Day Strength Challenge is designed to help anyone and everyone feel stronger at any age.

1 Begin by buying one bottle of MyoHealth Vegan Lemonade Powder. When you do, you'll receive a **second bottle-FREE! PLUS:** A free copy of Robert Wolfe, PhD's book *The Building Blocks of Life*, and a handy shaker bottle.*

2 Double your dosage (that's two servings a day) for the first 30 days to build up your body and prove the power of MyoHealth.

3 Live it, love it and share your lifestyle improvement story. After you complete the 30-Day Challenge, send us your story and get a **FREE MyoHealth t-shirt**. Visit MyoHealthStories.com for more information.

BUY ONE MYOHEALTH ESSENTIAL AMINO ACID COMPLEX VEGAN LEMONADE POWDER

RECEIVE ONE FREE!*

THAT'S OVER $60 IN PRODUCT AND MATERIALS FREE!

CALL TOLL-FREE
1-800-991-7116
OR VISIT
MYOHEALTH.COM

*Special offer good to new and existing customers one time only.